D1304178

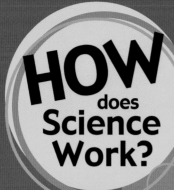

HOW does Science Work?

Exploring Sound

Carol Ballard

PowerKiDS press.

New York

Published in 2008 by The Rosen Publishing Group, Inc.
29 East 21st Street, New York, NY 10010

First Edition

Commissioning Editor: Vicky Brooker
Editors: Laura Milne, Camilla Lloyd
Senior Design Manager: Rosamund Saunders
Design and artwork: Peta Phipps
Commissioned Photography: Philip Wilkins
Consultant: Dr. Peter Burrows
Series Consultant: Sally Hewitt
Artwork p.11: Peter Bull

Library of Congress Cataloging-in-Publication Data

Ballard, Carol.
  Exploring sound / Carol Ballard.
    p. cm. — (How does science work?)
  Includes index.
  ISBN 978-1-4042-4279-1 (library binding)
  1. Sound—Juvenile literature. I. Title.
  QC225.5.B345 2008
  534—dc22
                                    2007032251

Manufactured in China

Acknowledgements:

Cover photograph: Drumming on Snare Drum, Louie Psihoyos/Getty Images

Photo credits: Austin Brown/Getty Images 4, Richard Price/Getty Images 6, Jonathan Cavendish/Corbis 7, Lothar Lenz/Corbis 12, Richard Kolker/Getty Images 13, Phillip Colla/Ecoscene 14, Chris Shinn/Getty Images 16, Conrad Zobel/Corbis 18, LA/CABYVALENCE/Photolibrary 19, John Birdsall 20, Lemoine/PhotoLibrary 21, Arne Hodalic/Corbis 22, Merlin Tuttle/Science Photo Library 23, Louie Psihoyos/Corbis 26.

The author and publisher would like to thank the models Kodie Briggs, Dylan Chen, Jessica Li Murphy, and Isabelle Li Murphy.

Contents

Sound 4

Where does sound come from? 6

How does sound travel? 8

How do we hear? 10

Why do we have two ears? 12

What can sound travel through? 14

Blocking sounds 16

Looking after your ears 18

Being deaf 20

What is an echo? 22

High or low? 24

Loud or quiet? 26

Measuring sound 28

Glossary 30

Further information and Web Sites 31

Index 32

Words in **bold** can be found in the glossary on p.30

Sound

We live in a world filled with many different sounds. There are sounds around us all the time. You might think you are in a quiet room, but try shutting your eyes and listening carefully. How many different sounds can you hear?

Some sounds, such as shouting, are very loud. Other sounds, such as a whisper, are very quiet. Some sounds, such as foghorns, are very low. Other sounds, such as whistles, are very high.

An airplane taking off is noisy. →

Sounds are very useful. Whenever we talk, laugh, shout, cry, or whisper, we are using sounds to communicate with each other. Sounds such as fire alarms and police sirens warn us of danger. Musical instruments make sounds that we enjoy listening to. We rely on the sounds of doorbells, alarm clocks,and telephones every day.

Sounds help us to communicate with each other.

Where does sound come from?

Anything that makes a sound is called a **sound source**. Some sounds are made by the world in which we live. These are **natural** sound sources, such as the clap of thunder in a storm and the rustling of leaves in a breeze.

Waves crashing on rocks are a natural sound source.

Some sound sources are made by living things. These are **living** sound sources, such as dogs barking and birds singing.

Some sounds come from machines. These are **artificial** or **man-made** sound sources. Radios, alarm bells, and electric drills are all artificial sound sources.

⬆ Cats make a purring sound when they are petted.

How does sound travel?

Sound travels away from a sound source in all directions. A sound is made when a sound source **vibrates**. This means that it makes tiny, very fast backward and forward movements. As it does so, it makes the air around it vibrate, too.

Guitar strings vibrate when they are plucked.

The vibrations travel through the air, spreading out in wider and wider circles. These vibrations in the air are called **sound waves**.

TRY THIS! Make waves

1 You will need a bowl and some modeling clay.

2 Get your bowl and fill it halfway with water.

3 Drop the small ball of modeling clay into the middle.

4 Watch what happens to the surface of the water.

You should find that ripples spread out in wider and wider circles, away from the place where you dropped the modeling clay. Sound waves travel through the air in the same way as the ripples travel through the water.

Wow!

Sound travels at an amazing speed—about 1,129 feet (344 meters) every second through the air!

How do we hear?

We use our ears to hear sounds around us. When you look at your ears in a mirror, you see two flaps, one on each side of your head. These flaps are the outside part of your ear. The rest of each ear is hidden inside your head. Sound waves that enter our ears allow us to hear.

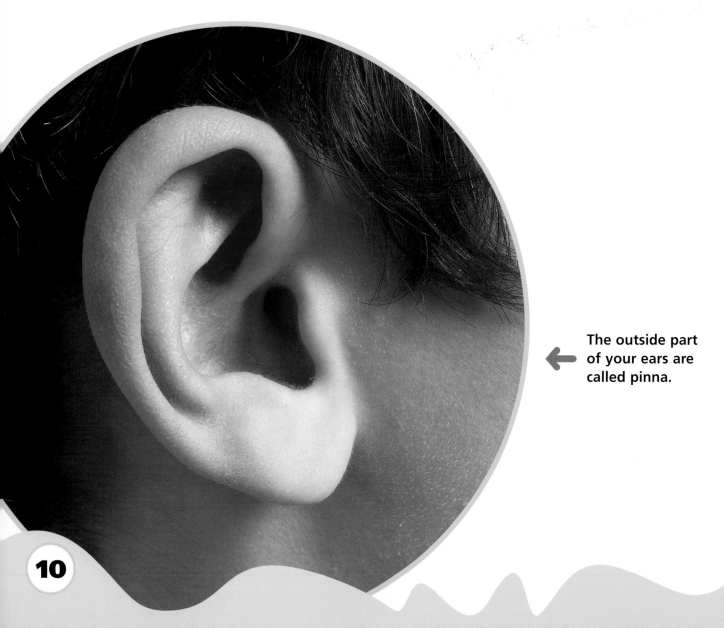

The outside part of your ears are called pinna.

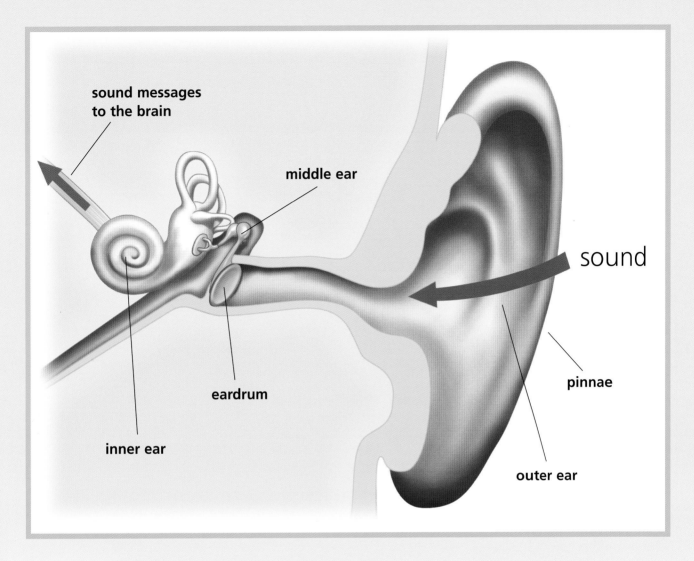

sound messages
to the brain

middle ear

sound

pinnae

eardrum

inner ear

outer ear

Sound waves travel into the **pinna** and through the outer ear. At the end of the **outer ear** is a thin layer of skin called the **eardrum**. The sound waves make the eardrum vibrate. The vibrations travel through the **middle ear**. Then vibrations travel along a spiral in the **inner ear**. The vibrations then connect to the **nerves**, which send messages to the brain. The brain works out what you are hearing.

A diagram of the ear.

Why do we have two ears?

Having two ears helps us to pinpoint which direction a sound is coming from. If one ear points toward a sound source, the sound waves will reach it just before they reach the other ear. The brain will get a message from that ear before it gets the message from the other ear. This helps your brain to work out which direction the sound came from.

Many animals can move their ears. This means that they can pinpoint the direction a sound is coming from much more accurately than humans.

Dogs have good hearing because they can move their ears. →

TRY THIS! Pinpoint sounds

1 Cover your eyes with a blindfold or scarf.

2 Ask some friends to clap one by one in different positions around you.

3 See if you can tell which direction the sound is coming from.

4 Try and guess who is closest to you.

You could probably pinpoint the direction of the sounds. Animals, such as dogs, have much better hearing and can tell where sounds come from more accurately.

If your eyes are covered and you cannot see where the sound is coming from, you can use your ears.

Your ears help you to tell which direction the sound is coming from.

Can you tell where a sound comes from?

What can sound travel through?

Sound needs something to travel through. It can travel through air and other gases. It can travel through water and other liquids. Sound can travel through solid materials, too. Sound cannot travel through an empty space. There is no sound on the Moon, because there is no air for sound to travel through.

Dolphins can communicate with each other, because they make sounds that travel through water. →

TRY THIS! Make a string telephone

1 You will need some scissors, two plastic cups, and some string.

2 Make a small hole in the bottom of each of the plastic cups.

3 Thread a piece of string through each hole.

4 Tie a knot in each end of the string so that the cups are linked together.

5 Give one cup to a friend. Walk away until the string is tight.

6 Speak into your cup —your friend should hear your voice by holding the cup to their ear.

This is a string telephone. It works because sound travels along the string.

Be careful using scissors

↑ **You can use a string telephone to send messages.**

Blocking sounds

Sounds can travel more easily through some materials than others. Sounds can travel through air, liquids, and through hard materials, such as metal and wood.

Sound travels less easily through soft, squashy materials, such as foam and thick fabrics. Materials that sound cannot travel easily through can be used to block sound. They are called **sound insulators**.

This man works in a very noisy place, so he wears ear protectors to stop his ears from being hurt. →

TRY THIS! Block sounds

1 Find some different materials, such as sponges, wood, and cloth—you will need two pieces of each material.

2 Ask a friend to cover their ears with the first material.

3 Make a sound. How well can your friend hear you?

4 Test each of your materials to see which ones sound travels through easily and which ones sound doesn't travel through as well.

You should find that the softer and thicker the material, the better they are at blocking the sound.

Looking after your ears

Ears are delicate and should be looked after properly so that they work well throughout our lives. Loud noises are especially bad for ears and can cause permanent harm.

Turning the **volume** down on personal stereos, and avoiding using them for long periods of time, can help your ears. People who work in very noisy places or use noisy machinery should wear ear protectors to prevent damage to their ears.

Rock concerts can be very loud and damage your ears. →

Ears need to be kept clean. This reduces the buildup of dirt and the chance of infections. Earache is a common problem and you may need to visit a doctor if your ears hurt. By looking inside the ear with an **otoscope**, the doctor can see what is causing the earache. Medicine can often cure the problem.

Doctors can examine your ears using an instrument called an otoscope.

Being deaf

People who are **deaf** and **hard of hearing** cannot hear properly. They need special ways of coping with everyday life. These people use sign language to communicate.

People who are deaf and hard of hearing cannot hear things such as alarm clocks, fire alarms, and television programs. A flashing light can replace the ringing or alarm sound, so deaf people can see them instead. Subtitles of conversations in movies and on television help them understand.

 Sign language is a way of talking by using hand gestures.

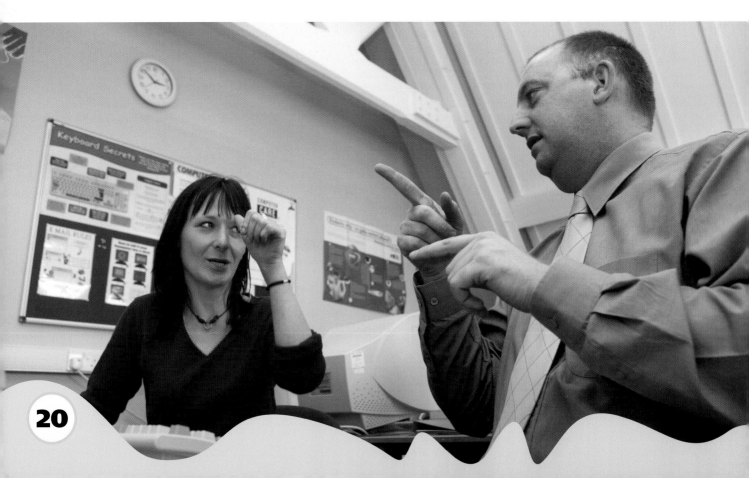

Many deaf and hard of hearing people wear **hearing aids** that make sounds louder and easier to hear.

They can often lipread. They watch the shapes and movements of a speaker's lips and work out the sounds that are spoken.

Hearing dogs can help, too. They are trained to alert their owners when they hear important sounds, such as telephones and doorbells.

Hearing aids are tiny microphones that fit inside or behind the ear.

What is an echo?

An **echo** is a sound that bounces back from a long way away. If you shout loudly in a place such as a cave, you might hear your voice again several times.

The sound of your voice travels through the cave. When it hits the hard stone of the cave wall, it bounces off of it. When it reaches your ears, you hear it again. This is an echo.

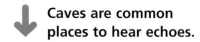

Caves are common places to hear echoes.

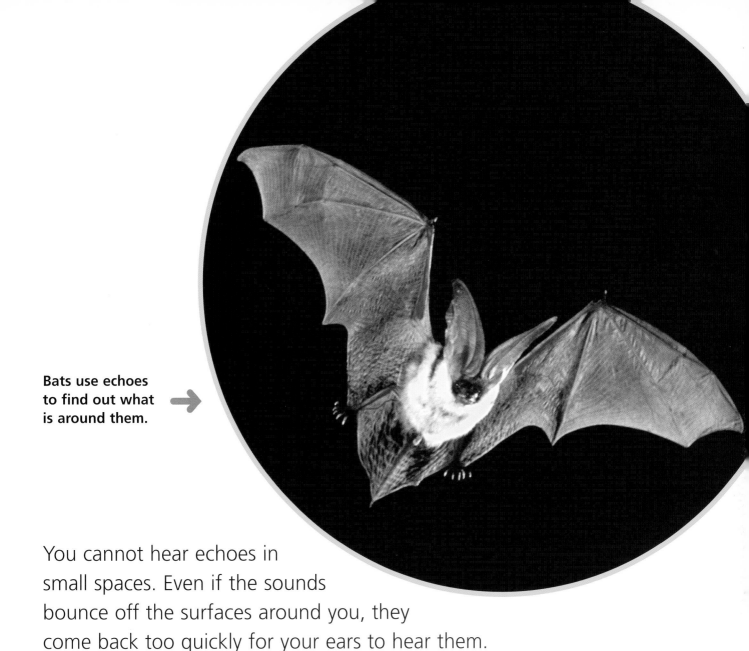

Bats use echoes to find out what is around them. ➡

You cannot hear echoes in small spaces. Even if the sounds bounce off the surfaces around you, they come back too quickly for your ears to hear them.

Some animals use echoes. Bats send out a series of tiny squeaks, which bounce back from things around them. By detecting the echoes, the bats can fly without bumping into things. They are also able to find their **prey** in this way. Whales and dolphins also use echoes to help them find their way.

High or low?

How high or low a sound is, is called its **pitch**. If a pebble is dropped into water, ripples will spread out in all directions. If the ripples are very close together, they will seem to move very fast and there will be a lot of them. If the ripples are farther apart, they will seem to move slowly and there will be fewer of them.

Sound waves act in the same way. The differences in the waves make the different pitches of the sound. Fast, close sound waves make high-pitched sounds. Slow, far-apart sound waves make low-pitched sounds.

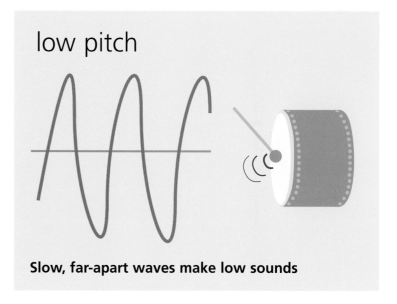

low pitch

Slow, far-apart waves make low sounds

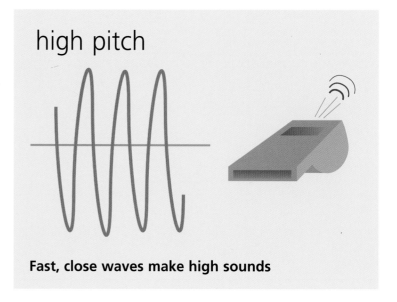

high pitch

Fast, close waves make high sounds

TRY THIS! Make different pitches

1 You will need several glass bottles or drinking glasses.

2 Pour different amounts of water into each.

3 Gently tap each glass or bottle with a metal spoon.

4 Try to put your bottles in order of pitch— start with the lowest on the left.

You should find that the more water there is in a bottle, the lower the sound it makes.

This is because the more water there is, the slower the sound waves travel.

Wow!

Elephants can hear sounds that are much too low for humans to hear!

Loud or quiet?

How loud or quiet a sound is, is called its volume. Sound waves can be very big or very small. Different sizes of sound waves make sounds of different volumes. Big sound waves make loud sounds and small sound waves make quiet sounds.

If you bang a drum hard, large sound waves are made.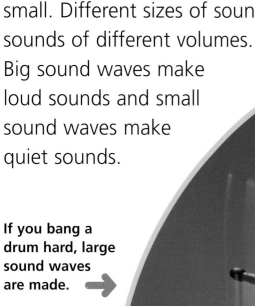

TRY THIS! Making vibrations

1 You will need some plastic wrap, a bowl, and some rice.

2 Stretch a piece of plastic wrap tightly over the top of a bowl.

3 Put a few grains of rice on top of the plastic wrap.

4 Tap the plastic wrap gently with one finger. Listen to the sound and watch the rice.

5 Tap the plastic wrap harder, and then even harder.

You should find that the harder you tap the plastic wrap, the louder the sound. The louder the sound, the bigger the vibrations, and so the higher the rice grains will jump.

Measuring sound

The volume of a sound is measured in units called **decibels**. The volume of a sound can be measured using an instrument called a **decibel meter**.

Some sounds are so quiet that humans cannot hear them. The quietest sound we can hear is called the **hearing threshold**, at 0 decibels. When sounds are very loud, they are close to the **threshold of pain** and then they can hurt our ears!

This is a decibel scale showing the volumes of sounds.

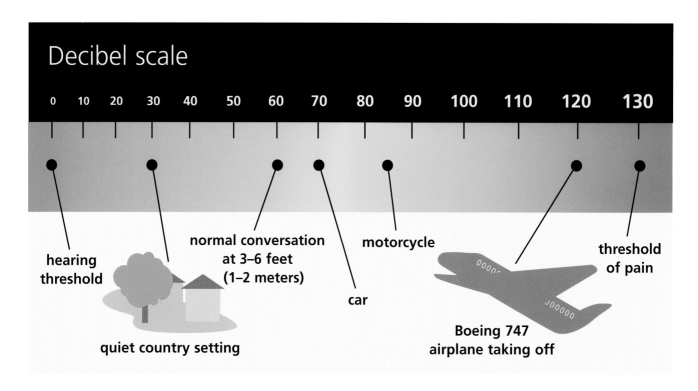

Decibel scale

0 10 20 30 40 50 60 70 80 90 100 110 120 130

hearing threshold

normal conversation at 3–6 feet (1–2 meters)

motorcycle

threshold of pain

car

quiet country setting

Boeing 747 airplane taking off

Sounds can be a nuisance and disturb people around them. Places that are very noisy can be asked to lower the sound levels. Sounds we do not want to hear are called **noise pollution**.

If you try banging a saucepan with a spoon, you will see that you can make a loud sound. This means that the sound waves are large.

Banging a saucepan with a wooden spoon makes a very loud sound. →

Glossary

artificial made by machinery or equipment

deaf unable to hear properly

decibels units for the volumes of a sounds

decibel meter an instrument that measures the volume of a sound

eardrum thin layer of skin inside the ear

echo a sound bouncing back

hard of hearing unable to hear properly

hearing aid device to help deaf people hear

hearing threshold the point where you can hear the quietest sounds

inner ear the part of the ear deep inside the head

living alive

man-made made by people

middle ear part of the ear between outer ear and inner ear

natural to do with the world around us

nerves parts of the body that carry messages

noise pollution unwanted noise

otoscope a medical instrument for examining the ear

outer ear outermost part of the ear

pinna part of the ear you can see at the side of the head

pitch how high or low a sound is

prey animal that is food for another animal

sound insulators materials that block sounds

sound source anything that makes a sound

sound waves vibration that carries a sound

threshold of pain the point where sounds become painful to your ears

vibrate move backward and forward rapidly

volume how loud or quiet a sound is

Further information

Books to read

Light and Sound (Science Files) by Chris Oxlade (Hodder Wayland, 2005)

Sound (Discovering Science) by Rebecca Hunter (Raintree, 2000)

Sound and Hearing (Start-up Science) by Claire Llewellyn (Evans Brothers, 2004)

Sound, From Whisper to Rock Band (Science Answers) by Chris Cooper (Heinemann Library, 2003)

Sound: Listen Up! (Science in Your Life) by Ben Craven (Raintree, 2005)

Web sites to visit

Web Sites

Due to the changing nature of Internet links, PowerKids Press has developed an online list of Web sites related to the subject of this book. This site is regularly updated. Please use this link to access this list: www.powerkidslinks.com/hdsw/sound

CD Roms to explore

Eyewitness Encyclopedia of Science, Global Software Publishing

I Love Science!, Global Software Publishing

My First Amazing Science Explorer, Global Software Publishing

Index

airplanes 4, 16, 28
alarm clocks 5, 20
animals 7, 12, 13, 14,
    21, 23, 25
artificial sound sources 4,
    5, 7, 16, 18, 24, 26, 28

bats 23

deaf 20–21
decibels 28
decibel meters 28
doctors 19
doorbells 5

earaches 19
ear protectors 16, 17, 18
ears 10–11, 12, 13,
    18–19, 22
echos 22-23

fire alarms 5, 20
foghorns 4

gases 14

hard of hearing 20–21
hearing 4, 5, 10–11,
    12–13, 15, 17, 20, 21,
    22, 28, 29
hearing aids 21
high 24–25

lipreading 21
liquids 14, 25
low 24–25

materials 17
microphones 21
Moon 14
musical instruments 5, 8,
    18, 24, 26

natural sound sources
    5, 6
noise pollution 29

otoscope 19

pitch 24–25
police sirens 5

prey 23

rock concerts 18

sign language 20
shouting 4
sound insulators 16,
    17, 18
sound sources 6–7, 1
    13
sound waves 8, 9, 1
    12, 24, 25, 26, 29
solids 14
speed of sound 9

telephones 5, 15, 21
thunder 6

vibrations 8, 11, 27
volume 18, 26–27,
    28–29

whispers 4, 5